THE EFFECT OF ICT ON THE PERFORMANCE OF SMALL AND MEDIUM SCALE ENTERPRISES

BY

ADEBIYI EZEKIEL OLUWATOBILOBA

MAC/2012/006

A RESEARCH PROJECT SUBMITTED TO THE DEPARTMENT OF MANAGEMENT AND ACCOUNTING, OBAFEMI AWOLOWO UNIVERESITY, ILE – IFE, OSUN STATE ,NIGERIA.

IN PARTIAL FULFILMENT OF THE REQUIREMENT S FOR THE AWARD OF BACHELOR OF SCIENCE HONOURS DEGREE IN MANAGEMENT AND ACCOUNTING

2017

CERTIFICATION

I certify that the research work entitled "The effects of ICT on the performance of SMEs" was carried out by ADEBIYI EZEKIEL OLUWATOBILOBA MAC/2012/006 in the department of Management and Accounting under my supervision.

_____   _____

DR. O. O OLASANMI   date

_____   _____

PROF. D. O ELUMILADE   date

DEDICATION

This research project is dedicated to my foster mother Alhaja Silifat Mopelola Adebiyi for her support and encouragement throughout my degree program. I salute you ma.

ACKNOWLEDGEMENT

First and foremost, I give thanks to the Almighty God the creator of heaven and earth for his guidance over my life. I give him all the glory for he is God alone. From the commencement of the research work to the end of the research work, he has been there for me.

My appreciation goes to my project supervisor, DR O.O Olasanmi for her guidance and her supervisory role in the conduct of this research work. Your corrections and guidance has really impacted my life and on my research work.

My appreciation also goes to father, Gabriel Adebiyi for contributing immensely to this research work. I appreciate your support financially for the success of this project. I also appreciate those who have contributed their quota for the success of the project like Olubunmi Olumoyi and Folakemi Famoyin Abi. The Lord will enrich your purse in Jesus name.

TABLE OF CONTENTS

Title   i

Certification   ii

Acknowledgement   iii

Dedication   iv

Table of Contents   v

Abstracts   viii

List of tables   ix

Chapter 1   1

INTRODUCTION   1

1.1 Background to the Study   1

1.2 Statement of the Problem   4

1.3 Research Questions   4

1.4 Research Objectives   4

1.5 Research Hypothesis   4

1.6 Operational definition of Terms   4

1.7 Scope of the Study   5

Chapter 2   6

LITERATURE REVIEW   6

2.1 Empirical Review   6

2.2 Theoretical Review   8

2.3 Conceptual Review   10

     2.3.1 Small and Medium Enterprises   10

     2.3.2 Profile of SMEs   11

     2.3.2 Type of SMEs   11

     2.3.4 Life Cycle of Small and Medium Enterprises   12

     2.3.5 Contribution of Small Scale firms   12

     2.3.6 Problem affecting Small and Medium Enterprises   13

     2.3.7 Information Communication and Technology   14

     2.3.8 ICT diffusion in developing countries   15

     2.3.9 ICT adoption and firm performance   15

     2.3.10 Role of ICT in SMEs   17

     2.3.11 Barriers to Information Technology adoption by SMEs   17

     2.3.12 ICT in Nigeria   17

Chapter 3   19

METHODOLOGY   19

3.1 Area of the Study   19

3.2 Research design   19

3.3 Population, Sample Size and Sampling Techniques   20

3.4 Data Collection and Sources   20

3.5 Research Instrument   20

3.6 Measurement of Variables 20

3.7 Data Analysis Techniques 21

Chapter 4 22

PRESENTATION, ANALYSIS AND INTERPRETATION OF DATA 22

4.1 Demographic Characteristics 22

4.2 Use of ICT equipment 24

4.3 Effects of ICT on SMEs performance 24

4.4 Factors Affecting ICT Usage 25

4.5 Mechanisms to enhance full Potentials of ICT 26

4.6 Test of Hypothesis 27

4.7 Limitations of the study 29

Chapter 5 30

Summary, Conclusion and Recommendations 30

5.1 Summary 30

5.2 Conclusion 30

5.3 Recommendations 31

Bibliography 32

Appendix

ABSTRACT

The purpose of the research work is to explore the effects of ICT on the performance of SMEs. The study also looked at the factors that affected the usage of ICT facilities and the ways and mechanisms to tap into the potentials of ICT to enhance the performance of SMEs.

The research was carried out in Ife Central Local Government. Primary data was used for the study. A simple random sampling with a sample size of 50 SMEs was selected from the area. Descriptive statistics such as frequency tables and inferential statistics such as correlation were used in the study.

It was discovered in the study that 80% of the respondents agreed that through the use of ICT, it will lead to the profitability of SMEs, 86% agreed that it will lead to increase in customer size,88% agreed that SMEs growth,86% and 90% agreed that it will lead improved service delivery and efficiency. It was also discovered in the study that 88% of the respondents agreed that poor power supply affects the usage of ICT facilities. 88%,82% and 90% of the respondents agreed that illiteracy, cost of ICT equipment and lack of maintenance of ICT facilities affect the usage ICT facilities among SMEs respectively. It was also discovered in the study that through training programs, government policy, awareness programs, cooperation among SMEs and consultancy service at low cost, it serves as a way of enhancing the full potentials of ICT as an enabler of socio economic development of SMEs as agreed by 90%,84%,92%,92 and %90% of the respondents respectively..

Hence based on the research, it is overviewed that through the use of ICT, it will make SMEs to be profitable. It will also make SMEs more efficient and it will make SMEs grow. It will also lead to improved service delivery and increase in customer size. It is also overviewed that power supply, lack of maintenance of ICT facilities, illiteracy and cost of ICT equipment affect the usage of ICT facilities. It is also overviewed that through training programs, government policies, cooperation among SMEs, consultancy services at a lower cost and awareness programs should be put in place to make sure that SMEs can tap into the potentials offered by ICT to enable socio economic development.

TABLES

Table List	Title	Page
Table 4.1	Demographic Characteristics	**23**
Table 4.2	Use of ICT equipment	**24**
Table 4.3	Effects of ICT on SMEs performance	**25**
Table 4.4	Factors affecting ICT Usage	**26**
Table 4.5	Mechanisms to enhance full potentials of ICT	**27**
Table 4.6	Correlation analysis	**28**

CHAPTER ONE

INTRODUCTION

1.1 BACKGROUND TO THE STUDY

In developed countries and developing countries Small and Medium Scale Enterprises play major role in industrial and economic growth. SMEs contribute significantly to the economic development of Nigeria. These contributions are remarkable as about 10% of the total manufacturing output and 70% of the industrial employment are by SMEs (Aina, 2007). The advent of ICT has shapen the way business is being conducted. The way and manner in which firms utilize ICT will go a long way in affecting their performance.

Information and Communication Technology (ICT) has become indispensable for small and medium enterprises globally, Nigeria inclusive. ICT capability is essential to participation and engagement in modern society. The development in telecommunications has impacted enormously on the applications of ICTs and their uses (Ladokun, Osunwole and Olaoye, 2013).The emergence of ICT has made the world a global village. The impact of ICT is being felt across all sectors and the business environment is not left out of its impact. Technology has changed the way all sectors have operated including the business sector and SMEs must tap into the opportunities provided by the ICT. The demand for ICT in the past decades is on the rise. The way in which ICT has been used to conduct businesses has changed over time. In the face of global competition firms looking for ways to survive have tapped into ICT and therefore have harnessed its benefits. The SME sector has a role it plays in economic development; it is used in tackling poverty and as a means of generating employment in developing countries (Hallberg, 2000). In the poor countries of the world i.e. the third world countries, SME is the sector in which most people who in those countries of the world operate on as a form of business. These countries run on SMEs as a means of survival. This sector contributes to the GDP of the national economies of the third world countries serving as the main source of employment opportunities. This has made the government of those countries and other organizations of those countries to recognize the need for that sector for overall economic development. This has made the government to come up with policies to support to its citizens who engage in SMEs to make sure that economic development is attained. The government of the third world countries came up with policies like tax incentives, technical assistance, from regulatory provisions to policy interventions, training and other types of business development services (O'Shea & Stevens, 1998).

SMEs grasp the ability to use to use ICT are more efficient. The Increasing trend in which companies have used networks by firms who are based on technology has given them a great advantage and hence SMEs need to tap into those advantages for improved performance. For an economy to be competitive, it requires the support and empowerment of the private sector and the SMEs. The use of ICT will cause changes in SMEs and other organizations.

Considering the potential of small and medium enterprises in economic development as well as their ability to innovate that has a large impact on long-term survival, that determines primary obstacles in the face the development of new technologies and acquisition as well as provide its appropriate tools and policies designed to overcome these challenges, have great importance.

Today revolution in information and communication technology has changed in a way that people and organizations are conducting their business. As a result of globalization and competitive demand for new technologies and innovations have laid the ground for ICTs to thrive in businesses and SMEs are not exempted in that. Globally, entrepreneurs and other business stakeholders are fully aware of the transformational impact of ICT for improvement both in the economic welfare and social welfare (Esselar, Stork, Ndiwalana and Dean-Swarray, 2007). Africa as a whole has the highest market for ICT and it also has the highest penetration of SMEs and therefore for SMEs who are operating in the African markets has the full possibility to lower the costs, improve the profitability and improving on the way in which they generate revenue by leveraging on the potentials available by communication technologies (Rufai, 2014). African technologies especially in Nigeria can grow their revenue when they invest their money in ICT.

While communication technologies could boost performance of SMEs in developing economy, prevailing socio-economic factors and circumstances within the districts where firms operate significantly shape such influence, and might account for varying outcomes in the business performance of different firms. Hence, developmental efforts should go beyond ensuring universal, equal access to ICTs, but also marshaling the machinery of government through policy initiatives to facilitate, enhance and support local capacity for effective usage of ICT tools to achieve socio-economic objective (Rufai, 2014). Big businesses have taken the opportunity of ICT to gain the edge over their competitors unlike the small and medium enterprises. There is strong evidence that ICT is the driver for economic growth and government all over are driving SMEs to adopt ICT (Akomea-Bonsu, 2012).

Today, organizations are utilizing ICT not only to reduce cost but to render improved customer service. The adoption of ICT will have significant positive consequences on SMEs and consequently on the economy of a nation. When an SME uses ICT, it can provide SMEs with information that is valuable, improved performance, a good relationship with consumers, increase in efficiency and cost control (Akomea-Bonsu, 2012).

E-commerce, according to Bansal and Sharma (2006), is rapidly transforming the way of business functions are performed, posing new challenges to the entrepreneurial profession. We are in a world driven by innovations. Through the innovation of ICT in this modern world, it has really turned the world upside down as the world never remains the same. Through the use of ICT, it has changed the way the world operates SMEs inclusive. Through the use of ICT businesses have evolved from the manual operations of biro, pencils, books and other paper operations to the electronic operations of mobile phones, computers, the internet and other electronic devices.

In a global world, the use of IT to increase performance is one of the challenges being faced by SMEs presently in developing country due to the lack of knowledge on the benefit of IT in their businesses. However, there is the need for changing roles of SMEs operation to meet the global challenges. In developing countries, ability to continuously

upgrade functions, processes and productive in Business becomes a matter not only of innovativeness but ultimately one of survival. IT is expected to improve SMEs operation performance in a form of transactional convenience, saving of time and quick transaction.

Emphasis on impact of information technology on increase in performance on SMEs can be considered as an issue of much apprehension to Entrepreneurs and practitioners in developing economy like Nigeria (Olusola and Oluwaseun, 2013).

So many techniques in ICT are springing up and large firms are assessing these techniques in ICT in their advantage in the face of competition. SMEs need to tap to these innovations for sustainable growth, profitability and efficiency. The impact of ICT will go a long way in boosting the corporate image of these SMEs.

1.2 STATEMENT OF THE PROBLEM

Technology evolves in a fast changing world. ICT has changed the way businesses operate. Large firms with huge capital base are utilizing ICT facilities to their advantage. SMEs are prominent in developing economies like Nigeria. The problem has been looked into by various authors overtime. When SMEs use ICT, it is to their advantage but there are problems that SMEs face that makes the business enterprises not able to use ICT effectively to achieve improved performance especially in developing economies like Nigeria. Problems like low power supply, illiteracy, cost of equipment, and lack of maintenance of ICT facilities affect ICT usage for improved performance. These factors are common among SMEs. SMEs need to access ICT facilities for improved performance. Through the use of ICT, SMEs can thrive well as in large corporations getting competitive advantage.

1.3 RESEARCH QUESTIONS

Through the investigation of ICT on SMEs performance, questions relating to the problem arises:

I.   Does ICT really contribute to SMEs performance?

II. Does poor power supply, illiteracy, cost of equipment, and lack of maintenance of ICT facilities affect ICT adoption?

1.4   RESEARCH OBJECTIVES

The broad objective is to access the effect of ICT on SMEs

The specific objectives are to:

i.   determine the factors that affect ICT usage

ii. suggest ways and mechanisms on how best to enhance the full potential of ICTs as an enabler of socio-economic development of SMEs.

iii. determine the effects of ICT on SMEs performance.

1.5 RESEARCH HYPOTHESIS

H_0: There is no relationship between ICT and SMEs performance.

1.6   OPERATIONAL DEFINITION OF TERMS

ICT: An acronym for Information, Communication and Technology.

SMEs: An acronym for Small and Medium Scale Enterprises.

ICT Investment: it is the resource spent on Information and Communication Technology facilities with the hope that benefits will accrue from the resources spent on the investment. It is also the expenditure on ICT facilities that will yield benefits in the future.

SME Performance: It is a measure of how well or how badly Small and Medium Enterprises function or work.

ICT adoption: the plan of using ICT facilities with the hope that one will benefit from the usage of the facilities.

1.7   SCOPE OF THE STUDY

The study focuses on Small and Medium Enterprises (SMEs) located in Ife particularly in Ife Central Local Government. Obtaining responses from respondents in such areas might be easy due to emerging markets in those areas and with the fact that the Nation's foremost university, Obafemi Awolowo University is also located in the area. However, in the course of the research and collecting responses from the respondents through the research instrument, one might come across some illiterates and those who do not understand the English Language which is the official language in such areas which might pose limitations for those who operates in the SMEs in those areas. But through the help of translation and interpretation to such respondents, responses will be got from them.

Costs of also administering research instrument might also pose limitations but the researcher will try to make sure that those costs that will be incurred in the administration of research instruments will be minimized and the desired number of sample size targeted by the researcher be achieved.

SMEs operating in Ife Central Local Government in which there is presence of Obafemi Awolowo University and the fact that there are emerging markets will help in getting information relating to the problem.

CHAPTER TWO

LITERATURE REVIEW

As the 21st century emerges, changes in technology have taken place. ICT has grown in leaps and bounds. ICT has brought the world together as one. Trends in ICT have changed the way the world operates. Later on, the business world tapped into ICT and as such benefited from ICT investment. The Small and medium sized Enterprises (SMEs) are thriving in developing countries. The SMEs are the backbone of developing economies. They are the businesses whose capital is not worth the capital of large companies. As ICT is peculiar to large companies, SMEs also need to use ICT to reach a level playing ground with the large companies. As SMEs are springing up in the developing countries and ICT is gaining a foothold in the world at large, authors have looked into this problem and have assessed how the ICT application and investment will go a long way in improving the performance of SMEs with a view to making sure that the SMEs are on a level ground with large scale corporations. There are related literatures which have tried to look into the problem and therefore need to be reviewed to access if any weaknesses and limitations if any in their work. This is a review of literature in which authors have in one way or the other contributed to knowledge as related to the problem.

2.1   EMPIRICAL REVIEW

There is also an empirical review of works done by various authors in which conclusions were made by them in their works.

In the works of Esselar et al (2007), sample sizes of 280 SMEs in 14 African countries were used in their works. It was discovered that through ICT in the form of ICT usage and expenditure had effect on SMEs performance in the form of profitability and labour productivity.

In the works of Rufai (2014) he used a sample size of 59 SMEs and he discovered that through the advent of ICT media (internet, mobile phone and personal computer), the SMEs were performing better than they were before the use of ICT media.

In the works of Akomea-Bonsu (2012) while examining the effect of ICT on SMEs performance in Kumasi, Ghana, he discovered that ICT through the use of internet has contributed to SMEs performance in the form of increase in sales, profit and output. He also carried out a survey through the sample size of 40 SMEs in Kumasi, Ghana.

In the work of Olusola and Oluwaseun (2013) a sample size of 200 SMEs was used. The research was carried out in Lagos State and it was discovered through their works that ICT has played a significant role in the performance of SMEs through increase in productivity and economic activities.

In the works of Ashrafi and Murtaza (2008) a sample size of 51 SMES was used. It was discovered that the use of ICT through the usage of internet connection, ICT staff, usage of enterprise software and the type of website affected SMEs performance through increase in market shares, growth of sales revenue and cutting of costs and expenses. The research work was conducted in Oman.

In the works of Onu et al (2014), a sample size of 70 employees of sachet water firms were used. It was discovered in their works that the use of ICT has increased output of manufacturing firms and had increased labour productivity.

In the works of Olise et al (2014), a sample size of 40 SMEs that ICT adoption is a function of capital input, turnover value, asset value and business experience.

In the works of Akanbi (2015), she used a sample size of 1000 SMEs in Lagos State. She discovered that ICT through the use of mobile banking, ATM and the internet have improved SMEs performance by increasing profitability, productivity, sales growth, turnover, service delivery, market expansion and SMEs growth.

In the works of Ladokun et al (2013) through a sample of 70 SMEs, it was discovered that infrastructure, government policies, management support, level of security, maintenance cost, skills and training and investment affect ICT usage.

In the works of Ojukwu (2006), he used a sample size of 40 SMEs and he discovered in his work that inadequate power supply, incessant fuel crises, telecommunication, currency devaluation and government policies affect ICT adoption.

In the works of Tunji Oluwafemi (2015) he used a sample size of 62 SMEs and he discovered that firms introduce ICT to change process and product, make their services more easily tradable, reduce inventories and open up new opportunities. The research work was carried out in Lagos, Nigeria.

In the works of Anga (2014), she used a sample size of 230 SMEs. She discovered that one of the major factors affecting Nigerian SMEs is ICT.

2.2   THEORETICAL REVIEW

There is also review of theoretical frameworks adopted by different authors in which they were able to base their conclusions upon. Some authors had a theoretical framework in which their findings were based. The theories are:

- **Social shaping theory**
- **Theory of collective action**
- **Integrationist impact theory**
- **Contingency Theory**
- **Technological Adoption Theory**

SOCIAL SHAPING THEORY

Adeniyi Rufai (2014) based his work on Social Shaping Theory. In his opinion, Social Shaping Theory states that there are sociological factors which will influence technology to be adopted in some areas. There must be a symbiotic relationship between technology and society. Existing social economic factors within the districts where the firms operate also influence the choice of communication technology and shaped its impact on business performance.

In Social shaping theory according to Rufai (2014) Social Context will determine the effect of ICT adoption on performance. In Social shaping Theory, there is a difference between technology and society. It is the society that determines technology to be adopted. The impact of ICT on SMEs performance varies in a sociological setting.

THEORY OF COLLECTIVE ACTION

Akande (2013) anchored his work on Theory of Collective Action. The theory of collective action states that groups which are small come together to achieve their aim. The group comes together with common interest to achieve their objectives. This is only feasible when a small group tries to overrun a large group. In theory of collective action, the minority group will try to overcome a majority group in other to fight for their interest. In a democratic setting, there is also a principle of majority rule. The opinion of the majority tries to suppress the opinion of a minority. In the theory of collective action, the minority rule will want to claim their right over the majority rule to make sure that their interests are acted upon. The theory of collective action tries to correct the flaws in a democratic setting. Cases in the theory of collective action includes, workers demanding for higher pay, militant groups demanding for a cause from the government and SMEs demanding for government attention. The theory of collective action states that a group enjoy the benefits that accrue to it but if the benefit does not the entirety of the group, the minority section will advocate for such benefits.

INTEGRATIONIST IMPACT THEORY

Ojukwu (2006) based his work on integrationist impact theory. Integrationist impact theory according to Ojukwu is the process of integrating the human input into the SMEs to achieve sustainable performance. He said that through the human input, there will be achievement using the ICT. He is of the opinion that human beings use ICT facilities for the development of their organization. A cursory look into the human capability according to him is beneficial to any organization.

CONTINGENCY THEORY

In the works of Anga (2014), she based her work on Contingency Theory. The theory states that there are internal and external factors that can affect an organization. There is no optimal method to systematize a firm and organizational structure of the company. A person who acts as a contingent leader will base its own style of leadership on either the internal situations or external situations. The theory states that there is no best approach of management. It states that different situations will warrant management style.

There are changes in situations such as changes in government policy, Change in environment, Customer demand and so on that will have impact on decisions taken by the management and approach by the management. Any decision taken by the management must be influenced by the environment in which the business operates. It calls for management decisions and approaches to be flexible.

TECHNOLOGICAL ADOPTION THEORY

Technology Adoption Theory was used by Olise et al (2014). This model states how users come to accept a new technology. The theory was propounded by Fred Davis. It stated that as technology evolves, users will be influenced to accept a new technology. The theory stated that the degree by which a person uses a technology will enhance job performance. The theory also states that it is the degree to which a person who uses a particular technology will be free from effort.

2.3   CONCEPTUAL REVIEW

2.3.1   Small and Medium Scale Enterprises

Ojo,(2004); contends that the "definition of small and medium scale enterprises varies according to context, author and countries". Small and medium scale enterprises are certainly not transnational company, multinational cooperation, publicly owned enterprises or large facility of any kind.

Ashrafi and Murtaza (2008) also defined SMEs as Businesses with less than ten employees as a Micro Enterprise, between ten and fifty as Small Enterprises, and between fifty to two hundred and fifty employees as Medium sized enterprises.

According to Gilaninia and Shahraki (2011) stated that as Small and Medium Enterprises (SMEs), comprise of institutes which have less than 250 employees, their annual turnover doesn`t exceed 50 million euros, and their annual balance sheet total is not more than 43 million euros in European countries.

The Small and Medium Sized Development Agency of Nigeria (SMEDAN) defines SMEs based on the following criteria: a micro enterprise as a business with less than 10 people with an annual turnover of less than ₦5,000,000.00; a small enterprise as a business with 10-49

people with an annual turnover of ₦5 to ₦49,000.000.00; and a medium enterprise

19

as a business with 50-199 people with an annual turnover of ₦50 to ₦499,000.000.00(€228,469,28).

SMEs statistical definition usually varies per country. However, most of the time the choice whether or not a company is an SME is based on the number of employees, value of assets or value of sales (Hallberg, 2000).

2.3.2 PROFILE OF SMES

According to Lali (2010) he stated that:

1. SMEs are flexible against changing of market and environment.

2. Activity and initiative of individuals in these firms is leads to immediate results.

3. Employees of these companies are highly motivated.

4. The initial capital required for these firms is limited.

5. Efficiency of capital is high in these firms.

6. These companies are the main driving forces of entrepreneurial development.

2.3.3 TYPES OF SMALL AND MEDIUM SCALE ENTERPRISE

According to Fasua (2006) he specified that there are types of Small and Medium Scale Enterprises (SMEs) which are:

i. **Micro/Cottage Industry:** That is an industry with total capital employed of not more than ₦15million working capital but excluding cost of land and a labour size of not more than 10 workers.

ii. **Small – Scale Industry:** An industry with total Capital employed of over ₦1.5million but not more than ₦50million including working capital but excluding cost of land and labour size of 11 – 100 workers.

iii. **Medium – Scale Industry:** An industry with a total capital employed of over ₦50million but not more than ₦200million including working capital but excluding cost of land, and or a labour size of 101 – 300 workers.

iv. **Large – Scale Industry:** An industry with a total capital employed of over ₦200million including working capital but excluding cost of land or a labour size of over 300 workers.

2.3.4 LIFE CYCLE OF SMALL AND MEDIUM ENTERPRISES

In general, the ideal life cycle of small and medium enterprises can be divided into four phases : start up, accelerated growth, stable growth and maturity.

i. Startup usually lasts for a period of one of three years during which the founder supervises the whole business activities that may be carried out by family members, friends or small number of employees.

ii. The Phase of Accelerated Growth usually lasts three to four years. During this period, the founder or a management expert handles management. At this point, a corporate organization is developed thereby leading to separation of ownership from management.

iii. The Stable Growth phase typically has duration of two to five years. During this period, management expertise and the corporate organization are divided into numerous departments and inflow of stable, long-term venture capital from corporate investors begins to appear (Oluwafemi, 2015).

2.3.5 CONTRIBUTIONS OF SMALL SCALE FIRMS

Small scale firms have made the following contributions according to Staley and Morse (1965):

i Raising the Efficiency of the Industrial System: Small scale industries have a lower capital or labour ration that is they are generally labour-intensive, thereby contributing towards the utilization of idle hands, thus using resources which would have been otherwise idle.

ii. Development of Entrepreneurs and Managers: Small scale industries are incubators for entrepreneurs and future manager of industry.

iii. Capital Formation (Human and Materials): Small scale industries make a significant contribution toward the creation of wealth and manpower development in the economy

iv. Capital Saving: They contribute towards generation of investible funds in the form of savings.

iv Employment: Small scale industries provides employment for no small percentage of the working population in any economy.

v Geographical Spread of Development: Because of their nature, many small scale industries are spread across the country. Thus, they aid rural development.

vi Socio-Political Development: By providing means of employment to a vast majority of the citizenry, small scale industries help to raise the standard of living of the people and help to arrest rural-urban migration so doing they reduce social problem which could otherwise threaten political stability.

vii Labour-Social Relations: As pointed out earlier, small scale industries serve as a training ground for young managers to develop managerial skills. This is equally so in labour relations, they also provide avenues for people to develop social relations.

viii National Enterprise: Some small scale enterprises contribute tremendously towards a nation's effort towards self-reliance. Some of the small-scale enterprises eventually grow to assume commanding heights in the economy. This role is of great significance for the economic and political security of a nation.

2.3.6 PROBLEMS AFFECTING SMALL & MEDIUM SCALE ENTERPRISES (SMES)

The greatest problems facing small and medium scale enterprise in Nigeria as put up by Oluwafemi (2015) are as follows:

i. **Finances**: In most cases, finance is the key problem in any industrial set up. Often this is due to lack of proper management or its misappropriation in adequate finance could strangle any business ventures. In Nigeria, the first idea for sources of funds to finance an enterprises would include personal savings ,contributions from friends and relations, credit financing from neighborhood, sale of personal/family properties, credit financing from commercial banks and loans from government. Furthermore, since most of the small business in developing countries in general are individuals or family business operating in a low income economy they are unable to generate sufficient funds through personal savings for the financing of huge capital equipment or the other fixed asset

ii. **Lack of managerial skills**: This is another militating force against small scale business. It should be stated that money for financing business is not the critical administration. This means that proper management of small–scale firms which entails bringing both human materials resource together to achieve set organization goals poses a problem to the small scale industrialist. Those who manage small-scale industries in Nigeria are largely illiterates or semi-illiterates who do not possess the knowledge of scientific management. The result is the management of small-scale firms in Nigeria is subjected to trial and error until disastrous consequences, for the owners of such industries. Power supply is an extremely vital input of small scale industries.

iii. **Poor communication facilities**: These are also some of the major problems confronting the small scale firm in Nigeria. These include the road, rail, water and our transport systems as well as provide direct services such as those of the doctor or the solicitor.

iv. **Limited market**: Due to generally low income per head in Nigeria, demand for goods and services is very low, large scale production is possibly only when there is available market for the commodity with limited market for their products, many firms in Nigeria still remain small in size.

v. **Lack of trust/the desire for independence**: Many businesses in Nigeria fear to combine with others due to lack of trust and also due to their desire for independence hence they are unable to finance a trust fund; they thus remain small.

vi. **Frequent change of government**: This creates fears in the mind of businessmen, because change of government, always affects their policies and attitude towards some industries. For instance, the new government might decide to wipe out some established industries.

vii. **Inadequate Infrastructural Development**: The Government has not done enough to create the best conducive environment for the striving of SSE, the problem of infrastructures ranges from shortage of water supply, inadequate transport systems, lack of electricity to solid waste management. Businesses have to provide expensive parallel infrastructure.

2.3.7   INFORMATION AND COMMUNICATION TECHNOLOGY (ICT)

The use of computers, peripheral devices connected to it and communication tools for collecting, processing, storage and dissemination of information is called information and communication technology. Information and communication technology is a term that is applied to any communication device or program, such as: radio, television, cellular phones, computers, software, hardware, networking, satellite systems and the like that is related numerous services, programs and services to them according to Gilaninia,Mousavian,Omidvari,Bakhshalipour and Eftekhari(2012). Malekian,(2010) viewed ICT as a specific concept and position and in a more accurate review of applications, such as information and communication technologies in education and health, libraries etc. and convergence between computer and communications. The most important feature of information and communication technology is a storage method, processing and access to information.

ICT refers to a wide range of computerized technologies. ICT is any technology that enables communication and the electronic capturing, processing and transmission of information (Ashrafi and Murtaza, 2008). These technologies include products and services such as desktop computers, laptops, handheld devices, wired or wireless intranet, business productivity software such as text editor and spreadsheet, enterprise software, data storage and security, network security and so on (Ashrafi and Murtaza, 2008). In Nigeria, commonly used ICTs include Internet, Personal Digital Assistants (PDAs), Automated Teller Machines (ATMs), mobile phones and smart cards (Apulu and Ige, 2011).

2.3.8   ICT DIFFUSION IN DEVELOPING COUNTRIES

Lal (2007) investigating adoption of ICT in Nigerian SMEs, found that one of the major factors inhibiting ICT diffusion and intensive utilization is poor physical infrastructure. In developing countries some of the ICT adoption challenges include legal and regulatory issues, weak ICT strategies, lack of Research& Development, excessive reliance on foreign technology and ongoing weaknesses in ICT implementation (Dutta and Curry 2003).

2.3.9   ICT ADOPTION AND FIRM PERFORMANCE

Despite the potential benefits of ICT and e-commerce, there is debate about whether and how their adoption improves firm performance. Use of and investment in ICT requires complementary investments in skills, organisation and innovation and investment and change entails risks and costs as well as bringing potential benefits. While many studies point to the possibility of market expansion as a major benefit for SMEs, larger businesses can also expand into areas in which SMEs dominated. Moreover, it is not easy for SMEs to implement and operate an on-line business, as this involves complementary costs for training and organisational changes as well as direct costs of investing in hardware and software solutions according to OECD (2004).

2.3.10 THE ROLE OF ICT IN SMES

In the present knowledge-based economy, it is important for SMEs to adopt processes that enable them to provide services that will bring about competitive advantage. ICT has a significant positive impact on organizational performance (Maldeni and Jayasena, 2009) and is vital to SMEs. ICT is known as a major catalyst and enabler of organizational change (Hazbo, 2008). Without the utilization of ICT, it may be impossible for modern SMEs to compete as ICT has a significant impact on SMEs operations and is claimed to be crucial for the survival and growth of economies in general (Berisha-Namani, 2009).

ICT provides opportunities for business transformations (Chibelushi, 2008) and provide SMEs the opportunity to conduct business anywhere (Jennex, 2004).

The European Commission (2008), states that SMEs could use ICT in order to grow and to become more innovative. Hence, there is a need to encourage the use of ICT in SMEs and address the high cost of ownership of ICT equipment since it can help to improve technical and managerial skills, making available e-business solutions for SMEs. Love et al (2004) ascertain that the use of ICT offers many benefits to SMEs at different levels (operational level, tactical level and strategic level). In Africa, the use of ICT is very recent as compared to countries like the UK and USA, which is at a better stage (Harindranath et al, 2008). Chacko and Harris (2005) state that there are two ways SMEs can benefits from ICT, first, SMEs can be the producers of ICT or second, SMEs can be users of ICT with the intention to increase productivity or improve communication for reaching new customers.

Chacko and Harris (2005) also stated that the use of ICT by SMEs depends on the benefits the ICT tools can bring to the business, which means its usage depends on the cost effectiveness. The ICTs adopted by SMEs serve as basic tools for their business communication such as using either mobile phones or fixed lines. For example, after SMEs adopt ICT tools, they also use personal computers (PC) with basic software installed. They can enjoy improved communication (with suppliers, customers or employees and so on) and meet information processing needs. Having Internet presence also enable SMEs to enjoy improved communication tools such as email, file sharing, creating websites, e-commerce, among others (Chacko and Harris, 2005).

Ongori (2009) states that the use of ICT would help change the way businesses operate in this era of globalization by changing business structures and increasing competition, creating competitive advantage for businesses and by changing business operations. For these reasons, SMEs must have an ability to compete and dynamically respond to rapidly changing markets using ICT.

2.3.10 BARRIERS TO INFORMATION TECHNOLOGY (IT) ADOPTION BY SMES

Ovia (2000) categorized internal and external barriers that impede adoption of IT by SMEs in a developing country. The internal barriers include owner manager characteristics, firm characteristics, cost and return on investment, and external barriers include: infrastructure, social, cultural, political, legal and regulatory.

Ojukwu (2006) reported that inadequate power supply, incessant fuel crises, telecommunication, currency devaluation and government policies are barriers which affect ICT adoption in SMEs.

Ladokun et al (2013) reported that infrastructure, government policies, management support, level of security, maintenance cost, skills and training and investment are barriers which affect ICT adoption in SMEs.

Ashrafi and Murtaza (2008) have identified monetary costs of implementation, time to implement ICT project, lack of necessary internal skills, uncertainty about return on investment, Lack of available information about relevant technologies, lack of top management support as barriers affecting ICT usage in SMEs.

OECD (2004) reports that common barriers include: unsuitability for the type of business; enabling factors (availability of ICT skills, qualified personnel, network infrastructure); cost factors (costs of ICT equipment and networks, software and re-organisation, and ongoing costs); and security and trust factors (security and reliability of e-commerce systems, uncertainty of payment methods, legal frameworks) are barriers affecting ICT usage in SMEs.

2.3.11 ICT IN NIGERIA

Nigeria has a population of about 150 million people; it is believed that it has one of the largest markets in Africa (Broadgroup, 2005). The GDP growth rate is estimated to be more than 6% per year. This gives opportunity for businesses to participate in various services. According to Asaolu (2006) stated that in recent times, computers are deployed to every sectors of the economy. These are noticeable with the improvement in computer processing, applications and tools developed on regular basis.

Akpan-Obong (2009) states that Nigeria is a major actor in the ICT sector in Africa. The author added that Nigeria could take the lead in the ICT sector in Africa as a result of its policy approach to growing ICTs and the active roles of the SMEs. Nigeria has so far recorded significant achievements in ICT utilization, and has recorded higher growth rates in the penetration and diffusion levels of ICT. Akpan-Obong (2009) further added that Nigeria could lead the way to harnessing ICT for socio-economic growth. Also there are prospects for SMEs development and economic growth with the application of ICT despite the poor state of infrastructure in the country. Hence, there is a need for Nigeria SMEs to utilize ICT.

CHAPTER THREE

METHODOLOGY

3.1   AREA OF THE STUDY

The area of the study in which the researcher wants to conduct is Ife Central Local Government. Ife Central Local Government is in Osun State in Nigeria. The local government was created along with the state in 1991 and 30 local government areas in the state were created in which Ife Central Local government is part of .

The local government (Ife Central) is part of the district that makes up Ile –Ife. Other parts of the district that make up Ile-Ife are Ife – East and Ife – North.

Ife Central Local Government has the Obafemi Awolowo University, one of the prestigious universities in Nigeria. Through the university, Ife – Central has attained a good reputation of itself over time. Through the development of businesses in the university, the development has spread to other parts of the local government.

Ife Central Local Government is gradually emerging as a point in which economic activities will thrive in the future.

3.2   RESEARCH DESIGN

The research design employed in the research is a descriptive research. In the research employed, the researcher looks into the knowledge of the variables under consideration. The design gives information about ICT and SMEs performance. The survey research method under the arm of the descriptive information gives inferences about the population from the sample. Through the survey research method we make generalizations about the population from the sample .

Through the survey research method, we make generalizations about the population from the sample. Sample to be used for the research will be generated through the survey research. The author will carry out a field survey.

3.3 POPULATION, SAMPLE SIZE AND SAMPLING TECHNIQUES

The population of the study is the SMEs located at Ife Central Local Government.The sample size of 50 SMEs will be selected from the SMEs in Ife Central Local Government. The sampling technique of simple random sampling will be used in conducting the research.

3.4 DATA COLLECTION AND SOURCES

The source of the data to be used for the research is the primary data. The primary data will be generated through the administration of questionnaires to the respondents. The collection of data will be done solely by the researcher. The researcher may need the service of enumerators if the researcher wants them. The duration of collecting the data may vary between a day to three days.

3.5 RESEARCH INSTRUMENT

The research instrument to be used is the questionnaire. The research instrument is divided into four sections. The first section relates to the business information. The second section relates to the usage of ICT infrastructure. The third section relates to ICT effects on SMEs performance while the fourth section relates to the factors which affect ICT adoption. The fifth section relates to suggestions on ways and mechanisms on how best to enhance the full potential of ICTs as an enabler of socio-economic development of SMEs.

The questions eliminate technicalities to make sure that the questions are understandable. The instrument is preferred so that information concerning the problem can be got from the respondents within a short period of time. The questionnaire will be subject to verification by the supervisor before the questionnaire is administered to the respondents.

3.6 MEASUREMENT OF VARIABLES

SMEs performance acts as the dependent variable while ICT serves as the independent variable.

ICT will be captured through ICT usage which is the use of the internet, computers and mobile phones.

Performance is captured through profitability, service delivery, growth, customer size and efficiency.

3.7   DATA ANALYSIS TECHNIQUES

In analyzing the data, the descriptive and inferential statistical tools will be used. The descriptive statistical tools to be used are the frequency distribution tables.

The inferential statistical tools to be used are the t statistical tests for significance and correlation test which is part of the OLS test. These will be used to test the hypothesis. The use of Statistical Packages for Social Sciences (SPSS) software which is a data analysis software will be used for the data analysis techniques.

CHAPTER FOUR

PRESENTATION, ANALYSIS AND INTERPRETATION OF DATA

The section presents information on the demographic characteristics and other results based on the objectives of the study and any other relevant analysis.

4.1 DEMOGRAPHIC CHARACTERISTICS

The demographic data of SMEs are presented in tables 1-5. Table 1 represents information on business registration.

The results in table 4.1 below indicated that 70% of SMEs in Ife Central Local Government were registered while 30% of SMEs in the area were not registered. This shows that most SMEs in Ife Central Local government are registered.

The results in table 4.1 below also indicated that 80% of SMEs sampled in the area have been in operating between 1-10 years while 16% of SMEs sampled in the area have been operating between 10-20 years as only 4% of SMEs operating in the area have been in operation between 20-30 years. This shows that most SMEs operating in Ife Central Local Government have been in business between 1-10 years.

The results also in table 4.1 below also showed the line of trade engaged by the SMEs in Ife Central Local Government. 16% of SMEs sampled are into wholesale trade as 40% of SMEs are into retail trade as 44% of SMEs sampled are rendering services. This shows that retailing and services are predominant in the area.

The results also in table 4.1 below shows that 80% of SMEs sampled in the area have been in operating between 1-10 years while 16% of SMEs sampled in the area have been operating between 10-20 years as only 4% of SMEs operating in the area have been in operation between 20-30 years. This shows that most SMEs operating in Ife Central Local Government have been in business between 1-10 years.

The table below also shows that 76% of SMEs in the area have employee size ranging from 1-20 employees. The results also shows that 8% and 2% of SMEs sampled in the area have employee size of 21-200 and 200 and above respectively. 14% of the SMEs sampled have no employees.

The table below shows that 76% of the SMEs in the area are in sole proprietorship while 24% of SMEs are into partnership business. This shows that Sole traders are predominant in Ife Central Local Government.

Table 4.1 Demographic Characteristics

		Frequency	Percent
Registration Information	Registered	35	70%
	Not registered	15	30%
	Total	**50**	**100%**
Line of Trade	Wholesale	8	16%
	Retail	20	40%
	Service	22	44%
	Total	**50**	**100%**
Years in Operation	1-10	40	80%
	10-20	8	16%
	20-30	2	4%
	Total	**50**	**100**
Employee Size	1-20	38	**76%**
	21-200	4	**8%**
	200 and above	1	**2%**
	None	7	**14%**
	Total	**50**	**100%**
Business Form	Sole Proprietor	38	76%
	Partnership	12	24%
	Total	**50**	**100**

Source: Field Survey

4.2 Use of ICT equipment

The tables below present the result of the usage of ICT equipment. The tables below indicate that 52% of SMEs use the internet while 48% of SMEs do not use the internet. The tables also indicate that 52% of SMEs in Ife Central Local Government use computers while 48% of SMEs do not use the computers. Mobile phone has the highest in terms of 98% of SMEs using it showing that most SMEs in Ife Central Local Government use the mobile phones as a medium of communication. 24% of the respondents use fixed line phones while 6% of SMEs use fax which is an outdated form of communication.

Table 4.2: Use of ICT equipment

Do you use any of this ICT equipment?	YES	NO
Internet	52%	48%
Computers	52%	48%
Mobile phones	98%	2%
Fixed line phones	24%	76%
Fax	6%	94%

4.3 Effects of ICT on SMEs performance`

In the table below, 80% of respondents (SMEs) agreed that the use of ICT, it will lead to SMEs profitability while 10% were uncertain and 10% disagreed. Hence ICT will lead to SMEs profitability. 86% of the respondents (SMEs) agreed that through the use of ICT, it will lead to improved service delivery while 2% were uncertain and 12% disagreed . Also,88% of respondents also agreed that through the use of ICT, it will lead to the growth of SMEs while 8% were uncertain and 4% disagreed. Hence ICT will lead to SMEs growth. 86% of the respondents (SMEs) were of the agreed that through the use of ICT, it will lead to increase in the size of customers while 14% of the respondents were uncertain. This indicates that through the use of ICT, it will lead to increase in customer size. 90% of respondents (SMES) agreed that through the use of ICT it will make the business more efficient. 6% of the respondents were uncertain while 4% disagreed that through the use of ICT, the business will be efficient. This signifies that through the use of ICT, it will make SMEs more efficient.

Table 4.3: Effects of ICT on SMEs performance.

S/N	QUESTIONS	SA	A	U	D	SD	TOTAL
1.	Through the use of ICT, it will lead to SMEs profitability	50% 25	30% 15	10% 5	8% 4	2% 1	100% 50
2.	Through the use of ICT, it will lead to improved service delivery.	44% 22	42% 21	2% 1	10% 5	2% 1	100% 50
3.	Through the use of ICT, it will lead to SMEs growth	34% 17	54% 27	8% 4	2% 1	2% 1	100% 50
4.	Through the use of ICT, it will lead to increase in customer size.	40% 20	46% 23	8% 4	6% 3	0% 0	100% 50
5.	Through the use of ICT, it will make SMEs more efficient.	42% 21	48% 24	6% 3	2% 1	2% 1	100% 50

Source: Field Survey

4.4   Factors affecting ICT usage

The table below shows that 88% of the respondents (SMEs) agreed that power supply affects ICT usage. 88% of respondents also agreed that Illiteracy affects ICT usage. 82% of respondents also agreed that cost of ICT equipment affect ICT usage while 90% of respondents also agreed that lack of maintenance of ICT facilities affect ICT usage. This indicates that cost of ICT equipment, illiteracy, lack of maintenance of ICT facilities and poor power supply affect the usage of ICT facilities.

Table 4.4: Factors affecting ICT Usage

QUESTIONS	SA	A	U	D	SD	TOTAL
Poor power supply affects ICT usage.	26	18	1	4	1	50
	52%	36%	2%	8%	2%	100%
Illiteracy affects ICT usage.	24	20	3	3	0	50
	48%	40%	6%	6%	0%	100%
Costs of ICT equipment affects ICT usage.	15	26	5	4	0	50
	30%	52%	10%	8%	0%	
Lack of maintenance of ICT facilities affects its usage	20	25	1	3	1	50
	40%	50%	2%	6%	2%	100%

Source: Field Survey

4.5   Mechanisms to enhance the full potentials of ICT

In the mechanisms to enhance the full potentials of ICT as an enabler for socio economic development of SMEs, 90% of the respondents agreed that there must be training programs to enhance ICT skills among SME owners. 84% of the respondents (SMEs) agreed that the government must have a policy to make sure that SMEs use ICT. 92% of the respondents agreed that there must be awareness programs for SMEs about ICT use while 92% of respondents agreed that there must be cooperation among SMEs to enhance ICT use. 90% of respondents agreed that there must be consultancy services and professional advice to SMEs at a lower cost to enhance ICT use. This shows that there must be training programs, government policy, awareness programs, consultancy services to SMEs at lower cost and cooperation among SMEs to make sure that they are put in place to enhance the full potentials of ICT as an enabler for socio economic development of SMEs.

Table 4.5: Mechanisms to enhance full potentials of ICT for development

QUESTIONS	SA	A	U	D	SD	TOTAL
There must be training programs to enhance ICT skills among SME owners.	23 46%	22 44%	3 6%	2 4%	0 0%	50 100%
Government must have a policy to enhance ICT use among SMEs.	15 30%	27 54%	5 10%	2 4%	1 2%	50 100%
There must be awareness programs for SMEs about ICT	15 30%	31 62%	2 4%	2 4%	0 0%	50 100%
There must be cooperation among SMEs to enhance ICT use.	16 32%	30 60%	2 4%	2 4%	0 0%	50 100%
There must be consultancy services and professional advice to SMEs at a lower cost.	16 32%	29 58%	4 8%	1 2%	0 0%	50 100%

4.6 Test of Hypothesis

H_0: There is no relationship between ICT and SMEs performance

Since the correlation values are positive which indicates a positive relationship of ICT on SMEs performance and such probability statistics is statistically significant at 0.000, we reject the null hypothesis that ICT has no relationship with SMEs performance and accept the alternative hypothesis that ICT has a relationship with SME performance.

The correlation analysis on the effects of ICT on SMEs performance is given in table 4.6 below as the correlation coefficient is indicated in the asterisk (*) which shows that it is significant at 0.000. The asterisk indicates that the correlation is statistically significant at 99% level of significance.

TABLE 4.6: CORRELATION ANALYSIS ON THE RELATIONSHIP BETWEEN ICT AND SME PERFORMANCE

			Through the use of ICT, it will lead to SMEs profitability	Through the use of ICT, it will lead to improved service delivery.	Through the use of ICT, it will lead to SMEs growth	Through the use of ICT, it will make SMEs more efficient	Through the use of ICT, it will lead to increase in customer size
Spearmans rho	**Through the use of ICT, it will lead to SMEs profitability**	Correlation Coefficient	1.000	0.507*	0.503*	0.673*	0.511*
		Sig.(2tailed)		0.000	0.000	0.000	0.000
		N	50	50	50	50	50
	Through the use of ICT, it will lead to improved service delivery.	Correlation Coefficient	0.507*	1.000	0.486*	0.527*	0.441*
		Sig.(2tailed)	0.000		0.000	0.000	0.001
		N	50	50	50	50	50
	Through the use of ICT, it will lead to	Correlation Coefficient	0.503*	0.486*	1.000	0.612*	0.656*
		Sig.(2tailed)	0.000	0.000		0.000	0.000
		N	50	50	50	50	50

SMEs growth.						
Through the use of ICT, it will make SMEs more efficient.	Correlation Coefficient Sig.(2tailed) N	0.673* 0.000 50	0.527* 0.000 50	0.612* 0.000 50	1.000 50	0.734* 0.000 50
Through the use of ICT, it will lead to increase in customer size	Correlation Coefficient Sig.(2tailed) N	0.511* 0.000 50	0.441* 0.001 50	0.656* 0.000 50	0.734* 0.000 50	1.000* 0.000 50

Source: Computations and output of SPSS 20 based on field survey

The effect of ICT on SMEs performance is strong. There is a strong relationship and so the null hypothesis is rejected.

4.7   Limitations of the Study

This study uses a sample size of 50 SMEs which was below what was expected from the research. Some of the owners of the SMEs encountered in the field survey refused to divulge information. Due to the time frame of the research and the refusal of some owners of these businesses to supply information, the sample of 50 SMEs had to be used.

CHAPTER FIVE

SUMMARY, CONCLUSION AND RECOMMENDATIONS

5.1   SUMMARY

The first question of the study was based if ICT really contributed to the performance of SMEs in Nigeria. The study discovered that through ICT use, SMEs will be profitable. It will lead to improved service delivery. Through the study it was also discovered that through ICT, it will make SMEs to grow, lead to increase in customer size and make SMEs function efficiently. The study area was Ife Central Local Government and it was discovered in the study that most SMEs operating in the area have been in business between 1-10 years and the business form is sole proprietorship.

The research question was also based in poor power supply, illiteracy, cost of equipment and maintenance of ICT facilities affect ICT usage and it was discovered in the findings that those factors affect the usage of ICT facilities.

In ways and mechanisms to enhance the full potentials of ICT as an enabler of socio economic development of SMEs, it was discovered in the study that there must be training programs to enhance ICT skills among SME owners. It was also discovered in the study that there must be awareness programs for SMEs about ICT use, government policy for SMEs to use ICT facilities, cooperation among SMEs to use ICT facilities and consultancy services at lower cost.

5.2   CONCLUSION

From the context of the study, a sound conclusion can be drawn that the use of ICT has a great impact in the performance of SMEs. Hence based on this research it is overviewed that through the use of ICT, it will lead to profitability of SMEs. In this research, it was also discovered that through the use of ICT, it will lead to increase in customer size, SMEs growth, improved service delivery and efficiency.

Furthermore, ICT has the potential to improve the core business of SMEs in every step of its business process, through the use of information technology and SMEs can gain from developing capabilities for managing information, intensive resources, enjoy reduced transaction costs, develop capacity for information gathering and dissemination of international scale and gain access to rapid flow of information.

A conclusion can also be drawn that there are factors which affect the usage of ICT facilities such as power supply, illiteracy, cost of ICT equipment and lack of maintenance of ICT facilities affect the usage of ICT facilities.

5.3   RECOMMENDATIONS

Driven by the findings in this research, SMEs in Nigeria has a long way to go for the sector to be most relevant, focused, productive enough and play the crucial role it is expected to in relation to contributing to the growth and development of the economy of Nigeria. There need to be training programs for SMEs as discovered in the research for them to tap the full potentials provided by ICT. There must also be availability of ICT consultancy services to SMEs at a lower cost.

The government also has a role to play to make sure that ICT is available for SMEs to use. The government needs to have a policy on ground for SMEs to use ICT facilities so as to tap into the potentials afforded by ICT. The following recommendations were made from the findings below:

i.   There should be training programs for SMEs about the use of ICT facilities

ii.   The factors affecting the use of ICT facilities especially power supply needs to be looked into by the government.

iii.   There should be provision of ICT facilities to SMEs at a lower cost as this can be done by government and corporate organizations.

iv.   There should be provision of consultancy services to SMEs at a lower cost.

v.   There must be awareness programs so that SMEs can be aware of the use of ICT facilities to promote development.

BIBLIOGRAPGHY

Adewoye, J.O. and Akanbi, T.A. The role of Information, Communication and Technology Investmentnon the profitability of Dmall and Medium Scale Enterprises. A Case of Sachet WaterCompanies in Oyo State, Nigeria. *Journal of Emerging Trends in Economies and Management Sciences.* 3(1), pp.64-71.

Aina, O. C. (2007). The role of SMEs in poverty alleviation in Nigeria. [Online] Available: http://www.journalanduse.org/Assets/Vol3%20Papers/JOURNAL%2010.pdf (13th December, 2009).

Akande, O. and Oluwaseun, Y. (2015). An appraisal of the impact of Information Technology on Nigerian Small and Mesium Scale Enterprises (SMEs) performance. *International Journal of Academic Research in Management* 2(4), pp.140-152.

Akanbi, T.A.(2015). An influential analysis of the impact of E-commerce in the Nigerian Small and Medium Scale Enterprises. *International Journal of application or innovation in Engineering and Management* (IJAEIM). 4, pp.63-68.

Akomea-Bonsu, C. (2012) The impact of Information Communication and Technology (ICT) on Small and NMedium Enterprises (SMEs) in the Kumasi Metropolis, Ghana, West Africa. *European Journal of Business and Management* 4(20), pp.152-158.

Anga, M. (2014) Determinants Of Small And Medium Scale Enterprises In Nigeria. *JORIND* 12(1), pg 140.

Ashrafi, R. and Murtaza, M.(2008).Use and impact of ICT on SMEs in Oman. *Electronic Journal of Information Systems Evaluation,* 11(3), pp.125-138.

Apulu, I. G and Ige, O. E. (2011). Are Nigeria SMEs Effectively Utilizing ICT? *International Journal of Business and Management* Vol. 6, No. 6, pp. 209.

Berisha-Namani, M. (2009). The role of information technology in small and medium sized enterprises in Kosova, Fulbright Academy Conference 2009, [Online] Available: http://www.fulbrightacademy.org/file_depot/0-10000000/20000-30000/21647/folder/82430/Berisha+Paper+IT+in+SMEs+in+Kosovo.pdf (15thJuly, 2009).

Chacko, J. G., and Harris, G. (2005). ICT and Small, Medium and Micro Enterprises in Asia Pacific – size does matter. *Information Technology for Development*, 12, 2, pp.175-177.

Chibelushi, C. (2008). Learning the hard way? Issues in the adoption of new technology in small technology oriented firms. *Education + Training*, 50(8/9), pp.725-736.

Esselar, S. Stork, C. Ndiwalana, A. and Dean-Swarray,M. (2007) ICT usage and its impact on the profitability of SMEs in 13 African Countries. 4(1), pp.87-100.

European Commission. (2008). Making SMEs more competitive [Online] Available: http://ec.europa.eu/enterprise/sme/competitive_en.htm (30th October, 2008).

Gilaninia, S. Mousavian,N. Bakhshalipour, A. Eftekhari, F. and Seighalani Z.F. (2012). The Role of ICT in Performance of Small and Medium Enterprises. *Interdisciplinary Journal Of Contemporary Research In Business*. 3(9), pp.833-839.

Jennex, M. E., Amoroso, D and Adelakun, O. (2004). E-commerce infrastructure success factors for small companies in developing economies. *Electronic commerce Research*, 4, pp. 263-286.

Maldeni, H. M. C. M., and Jayasena, S. (2009). Information and communication technology usage and bank branch performance. *The International Journal on Advances in ICT for Emerging Regions (ICTer)*, 2, 2, pp.29 – 37.

Ladokun, I.O. Osunwole, O.O. and Olaoye, B.O. (2013) Information and Communication Technology in Small and Medium Enterprises: Factors affecting the Adoption and use of ICT in Nigeria. *International Journal of Academic Research in Economics and Management* 2(6) pg 74

OECD (2004), *ICT, E-Business and SMEs*, Paris: OECD.

Onu, C. A. Olabode, I. O. and Fakunmoju, S. K (2014). Effect of Information Technology on SMEs Productivity and Growth. *European Journal of Humanities and Social Sciences* vol.32 no. 1 pg 1718

Olise, C. M. Anugbogu, T. U. Edoko,D. T. Okoli, I.M. (2014) Determinants of ICT adoption for improved SMEs performance in Anambra State, Nigeria. American *International Journal of Contemporary Research* 4(7), pp.163-176

Ojukwu, D. (2006) Achieving sustainable growth through the adoption of integrated business and information solutions. A case study of Nigerian Small and Medium Sized Enterprises. *Journal of Information technology impact.* 6(1), pp.47-60.

Rufai, A. (2014). The impact of communication technologies on the performance of SMEs in a developing economy: Nigeria as a case Study. *Electronic journal of Information Systemsin developing Countries.* 65(7), pp.1-22.

APPENDIX

DEPARTMENT OF MANAGEMENT AND ACCOUNTING

FACULTY OF ADMINISTRATION

OBAFEMI AWOLOWO UNIVERSITY, ILE IFE, NIGERIA

QUESTIONNAIRE

I am Adebiyi Ezekiel Oluwatobiloba, a final year student of the Department of Management and Accounting, currently carrying out a research on "Effects of Information, Communication and Technology (ICT) on SMEs performance". Please be assured that this is purely an academic exercise and you are enjoined to be as sincere as possible by giving relevant answers to the questions below using the instructions. Information given will be treated with the utmost level of confidentiality and anonymity. Thanks for your cooperation.

Instruction: Please tick [√] the correct box and complete the questionnaire as much as you can. Please respond to each question by ticking the appropriate box of your choice.

SECTION A: DEMOGRAPHIC DATA

1. Is the business registered?   (a) Yes [   ] (b) No [   ]

2. The business engages in which line of trade (a) Wholesale [   ] (b) Retail [   ] (c) Service [   ] (d) Manufacturing [   ]

3. Years in operation   (a) 1-10 years [   ]   (b)   10-20 years [   ]   (c) 20-30 years [   ]

4. Employee size   (a) 1-20 [   ]   (b) 21-200 [   ]
        (c) 200 and above [   ]   (d) none [   ]

5. The form of business   (a) Sole proprietorship [   ] (b) Partnership [   ]

SECTION B: This section contains questions regarding the familiarity with the use of ICT equipment. Please respond to the following questions by ticking in the boxes provided below.

Do you use any of this ICT equipment?	YES	NO

Internet		
Computers		
Mobile phones		
Fixed line phones		
Fax		

SECTION C: This section contains questions regarding the effects of ICT on the performance of your firm. Please respond to the following questions by ticking the correct answers below. You are to tick correctly in the spaces which indicates SA (Strongly Agree) , A (Agree), U (Uncertain), D (Disagree), SD (Strongly disagree).

S/N	QUESTIONS	SA	A	U	D	SD
1.	Through the use of ICT, it will lead to SMEs profitability					
2.	Through the use of ICT, it will lead to improved service delivery.					
3.	Through the use of ICT, it will lead to SMEs growth.					
4.	Through the use of ICT, it will lead to increase in customer size.					
5.	Through the use of ICT, it will make SMEs more efficient.					

SECTION D: This section contains questions that relates to factors that affect ICT usage in SMEs. Please respond to the following questions by ticking the correct answers below. You are to tick correctly in the spaces which indicates SA (Strongly Agree) , A (Agree), U (Uncertain), D (Disagree), SD (Strongly disagree).

S/N	QUESTIONS	SA	A	U	D	SD
1.	Poor power supply affects ICT usage.					
2.	Illiteracy affects ICT usage.					

S/N		SA	A	U	D	SD
3.	Costs of ICT equipment affects ICT usage					
4.	Lack of maintenance of ICT facilities affects its usage					

SECTION E: This section contains questions that relates with suggestions on ways and mechanisms on how best to enhance the full potential of ICTs as an enabler of socio-economic development of SMEs. You are to tick correctly in the spaces which indicates SA (Strongly Agree) , A (Agree), U (Uncertain), D (Disagree), SD (Strongly disagree).

S/N	QUESTIONS	SA	A	U	D	SD
1.	There must be training programs to enhance ICT skills among SME owners.					
2.	Government must have a policy to enhance ICT use among SMEs.					
3.	There must be awareness programs for SMEs about ICT					
4.	There must be cooperation among SMEs to enhance ICT use.					
5.	There must be consultancy services and professional advice to SMEs at a lower cost.					